The Beginner's Guide to Quantum Computing & Mechanics

Adam Smith

Published by A. Smith Media, 2022.

THE BEGINNER'S GUIDE TO QUANTUM COMPUTING & MECHANICS

First edition. December 12, 2022.

ISBN: 979-8215621479

Written by Adam Smith.

Table of Contents

"To all future quantum students and scientists, where are your shades? The future is bright."

- Adam Smith

Introduction

QUANTUM COMPUTING AND quantum mechanics are two fascinating and complex fields of study that have captured the imagination of scientists, researchers, and the general public alike. The principles of quantum mechanics, which describe the behavior of particles on the smallest of scales, are strange and counterintuitive and have challenged our understanding of the fundamental nature of the universe. Quantum computing, on the other hand, is a relatively new field that seeks to harness the principles of quantum mechanics to build powerful, highly efficient computers that can solve problems beyond classical computers' capabilities.

In this beginner's guide, we will briefly introduce the fundamental concepts of quantum mechanics and quantum computing. We will begin by exploring the basics of quantum mechanics, including the principles of superposition and entanglement and how they differ from classical mechanics. We will then delve into the world of quantum computing, including the basics of quantum algorithms and quantum circuits and how they can be used to solve complex problems.

Despite their complexity, quantum mechanics and quantum computing principles are not as difficult to understand as they may seem at first glance. With some background knowledge and basic mathematical concepts, anyone can begin to explore the fascinating world of quantum mechanics and quantum computing. So, let's dive in and learn more about these exciting fields of study!

Chapter 1: The history and development of quantum computing and its potential applications.

QUANTUM COMPUTING IS a relatively new field of study that focuses on developing computing technologies based on the principles of quantum mechanics. The idea of using quantum mechanics to create computers was first proposed in the 1980s, but it wasn't until the 1990s that researchers started to make significant progress in the field.

The fundamental difference between a classical computer and a quantum computer is the way in which they store and process information. Classical computers use bits, which are units of information that can be either a 0 or a 1. Quantum computers, on the other hand, use quantum bits, or qubits, which can be both a 0 and a 1 at the same time. This property, known as superposition, allows quantum computers to perform certain calculations much faster than classical computers.

The potential applications of quantum computing are vast and varied. For example, quantum computers could be used to solve complex mathematical problems, such as factoring large numbers, which are beyond classical computers' capabilities. This could have significant implications for cryptography, as many of the encryption algorithms used today rely on the difficulty of factoring large numbers.

Another potential application of quantum computing is in the field of chemistry. Quantum computers could be used to simulate the behavior of molecules, which could help chemists design new drugs and materials. In addition, quantum computers could be used to optimize complex systems, such as supply chains and traffic networks.

Despite the great potential of quantum computing, there are still many challenges to overcome before it becomes a practical technology. One of the biggest challenges is developing quantum algorithms that can take advantage of the unique properties of qubits. Additionally, quantum computers are extremely sensitive to outside influences, such as temperature and electromagnetic radiation, which can cause qubits to lose their quantum properties.

Overall, the field of quantum computing is still in its infancy, but it has the potential to revolutionize many different industries and fields of study. As research in this area continues to advance, we can expect to see more and more practical applications of quantum computing in the future.

Chapter 2: An introduction to the basics of quantum mechanics and its principles.

QUANTUM MECHANICS IS a branch of physics that deals with the behavior of particles on a very small scale, such as atoms and subatomic particles. It is based on the idea that energy, matter, and other physical quantities can be described in terms of certain fundamental units called quanta.

One of the key principles of quantum mechanics is the concept of wave-particle duality, which states that particles such as electrons can exhibit both wave-like and particle-like behavior. This is in contrast to classical mechanics, which describes the behavior of macroscopic objects in terms of their position, velocity, and other classical variables.

Another important principle of quantum mechanics is the uncertainty principle, which states that certain pairs of physical properties, such as position and momentum, cannot be measured with perfect precision at the same time. This is because the act of measuring one property will inevitably affect the other, due to the fundamental nature of reality at the quantum level.

Another key concept in quantum mechanics is the idea of superposition, which states that a particle can exist in multiple states simultaneously. This is what allows quantum computers to perform certain types of calculations much faster than classical computers.

Overall, quantum mechanics has profoundly impacted our understanding of the universe and has led to many significant technological advances, such as the development of transistors, lasers, and magnetic resonance imaging (MRI). It continues to be an active area of research, with many exciting developments on the horizon.

Chapter 3: The principles of quantum algorithms and their potential impact on fields such as cryptography and machine learning.

QUANTUM ALGORITHMS are a type of algorithm that make use of the principles of quantum mechanics to solve computational problems. Unlike classical algorithms, which are based on the principles of classical mechanics, quantum algorithms are able to exploit the unique properties of quantum systems, such as superposition and entanglement, to solve problems more efficiently. In recent years, there has been a great deal of interest in developing quantum algorithms for a wide range of applications, including cryptography and machine learning. In this essay, I will discuss the principles of quantum algorithms and their potential impact on these and other fields.

At the most basic level, a quantum algorithm is a sequence of quantum operations that are applied to a quantum system in order to solve a computational problem. The operations are chosen in such a way as to exploit the unique properties of quantum systems in order to solve the problem more efficiently than would be possible with a classical algorithm. For example, a quantum algorithm might make use of superposition, which allows a quantum system to exist in multiple states simultaneously, or entanglement, which allows two or more quantum systems to be correlated in ways that classical mechanics cannot explain.

One of the most well-known examples of a quantum algorithm is Shor's algorithm, which is used to factorize large numbers. In classical computing, factoring large numbers is a computationally difficult problem that is believed to be intractable for all but the largest classical computers. However, Shor's algorithm is able to factorize large numbers much more quickly using the principles of quantum mechanics. This has important implications for cryptography, as many modern cryptographic algorithms rely on the difficulty of factoring large numbers in order to ensure their security. By using Shor's algorithm, it may be possible to break these cryptographic algorithms, which could have serious consequences for the security of information transmitted over the internet.

Another area where quantum algorithms may have a significant impact is in the field of machine learning. Machine learning algorithms are used to analyze large amounts of data in order to make predictions or identify patterns. However, these algorithms are often limited by the amount of data they can process and the speed at which they can process it. Quantum algorithms, on the other hand, have the potential to process large amounts of data much more quickly and efficiently than classical algorithms. This could have major implications for the field of machine learning, allowing for more accurate and faster predictions and analyses of data.

Regarding their potential applications in cryptography and machine learning, quantum algorithms may also have important implications for other fields, such as drug design and materials science. In drug design, quantum algorithms could be used to predict the behavior of molecules and their interactions with proteins, allowing for the design of more effective and targeted drugs. In materials science, quantum algorithms could be used to simulate the behavior of materials at the atomic level, providing insights into their properties and potential applications.

The principles of quantum algorithms have the potential to revolutionize a wide range of fields, from cryptography and machine learning to drug design and materials science. While there are still many challenges to overcome in the development of these algorithms, their potential impact is vast and exciting. As more and more researchers begin to explore the possibilities of quantum algorithms, we may soon see them having a major impact on the world around us.

Chapter 4: The role of entanglement and superposition in quantum computing.

ENTANGLEMENT AND SUPERPOSITION are two fundamental concepts in quantum mechanics that have important applications in the field of quantum computing. Entanglement is a phenomenon in which two or more particles become strongly correlated, such that the state of one particle is dependent on the state of the other(s). Superposition, on the other hand, is the ability of a quantum system to exist in multiple states simultaneously. Together, these concepts allow for the creation of quantum states that can encode and manipulate information in ways that are not possible with classical computing.

In classical computing, information is represented by bits, which are binary units of information that can be either 0 or 1. Quantum computing, on the other hand, uses qubits, which are quantum mechanical systems that can exist in a superposition of 0 and 1. This allows for a much greater amount of information to be encoded in a single qubit, compared to a single bit.

Entanglement plays a crucial role in the manipulation of qubits in quantum computing. Creating entangled states between multiple qubits makes it possible to perform operations on the qubits that are not possible in classical computing. For example, entanglement can be used to create quantum gates, which are the building blocks of quantum circuits that can perform complex operations on qubits.

One of the key advantages of quantum computing is the ability to perform calculations on multiple qubits simultaneously. This is known as quantum parallelism, and it is made possible by the concept of superposition. In a quantum system, a qubit can be in a superposition of 0 and 1, allowing multiple calculations to be performed simultaneously. This can greatly increase the speed and efficiency of certain types of calculations, such as factoring large numbers or simulating quantum systems.

One of the most famous applications of entanglement and superposition in quantum computing is Shor's algorithm, which can be used to factorize large numbers quickly. This is a problem that is extremely difficult for classical computers, but can be solved efficiently on a quantum computer using entanglement and superposition. The algorithm uses a series of quantum gates to create entangled states between multiple qubits, which are then used to perform the necessary calculations.

Another important application of entanglement and superposition in quantum computing is quantum cryptography. This is a method of transmitting information securely, using the principles of quantum mechanics. In quantum cryptography, two parties can communicate using an entangled shared quantum state. This allows them to transmit information in a way that is secure, because any attempt to intercept the information would cause the entangled state to collapse, alerting the parties to the presence of an eavesdropper.

Entanglement and superposition are crucial concepts in the field of quantum computing. They allow for the creation of quantum states that can encode and manipulate information in ways that are not possible with classical computing. This opens up many potential applications, including faster and more efficient calculations, and secure communication using quantum cryptography.

Chapter 5: The challenges and limitations of current quantum computing technology include scalability and error correction issues.

QUANTUM COMPUTING IS a technology that has great potential in the way we process information and solve complex problems. In contrast to classical computing, which uses bits that can be in one of two states (0 or 1), quantum computing uses quantum bits or qubits that can be in a superposition of states. This allows quantum computers to perform certain calculations much faster than classical computers, making them a promising tool for tackling a wide range of challenges.

However, quantum computing technology is still in its infancy despite its potential and facing several challenges and limitations. One of the major challenges is scalability. Currently, the largest quantum computers consist of only a few hundred qubits, which is not nearly enough to tackle the most complex problems. To achieve the full potential of quantum computing, we will need to build quantum computers with thousands or even millions of qubits. This is a daunting task that requires significant advances in both hardware and software.

Another challenge is error correction. Quantum systems are highly sensitive to external perturbations, which can cause errors in the calculations performed by a quantum computer. To overcome this problem, researchers are working on ways to implement error correction algorithms that can detect and correct errors in the calculations. However, these algorithms are highly complex and require a large number of additional qubits, which can significantly reduce the computational power of the quantum computer.

Another limitation of current quantum computing technology is the difficulty in programming quantum computers. Unlike classical computers, which use well-established programming languages, quantum computers require specialized languages and algorithms that are still being developed. This means that only experts with extensive knowledge of quantum computing can program these machines, limiting their accessibility and potential applications.

Furthermore, quantum computers are also limited by their ability to perform only certain types of calculations. While they can perform some tasks much faster than classical computers, they are not universally superior. In fact, there are many problems that classical computers can solve more efficiently than quantum computers. This means that quantum computers are not a replacement for classical computers, but rather a complementary technology that can be used for specific types of calculations.

While quantum computing has the potential to revolutionize the way we process information and solve complex problems, it is still limited by several challenges and limitations. These include scalability, error correction, programming complexity, and the limited range of calculations that quantum computers can perform. Despite these challenges, researchers are making steady progress in overcoming these limitations and are working towards building larger and more powerful quantum computers that can be used to tackle a wider range of problems.

Chapter 6: The potential impact of quantum computing on fields such as chemistry, material science, and biology.

QUANTUM COMPUTING CAN drastically change many fields, including chemistry, material science, and biology. This is because quantum computers are able to perform calculations that are impossible or impractical for classical computers.

In chemistry, quantum computers could be used to simulate and predict the behavior of complex chemical systems. This could have a number of applications, such as the design of new drugs and materials, or the optimization of chemical reactions. For example, quantum computers could be used to study the behavior of enzymes, which are proteins that catalyze chemical reactions in the body. By simulating the behavior of these enzymes at a quantum level, researchers could better understand their function and potentially design new and more effective drugs.

In material science, quantum computers could be used to simulate the behavior of materials at the atomic level. This could allow researchers to design new materials with improved properties, such as increased strength or conductivity. For example, quantum computers could be used to study the behavior of superconductors, materials that can conduct electricity without resistance. By simulating the behavior of these materials at a quantum level, researchers could potentially design new and more efficient superconductors.

In biology, quantum computers could be used to simulate the behavior of biological systems at a quantum level. This could have a number of applications, such as the design of new drugs and the study of complex biological processes. For example, quantum computers could be used to study the behavior of proteins, which are essential for many biological processes. By simulating the behavior of proteins at a quantum level, researchers could better understand their function and potentially design new and more effective drugs.

The potential impact of quantum computing on fields such as chemistry, material science, and biology is enormous. By allowing researchers to simulate and predict the behavior of complex systems at a quantum level, quantum computers could lead to major advances in these fields and potentially profoundly impact society. However, it should be noted that quantum computers are still in the early stages of development, and it will likely be some time before their full potential is realized.

Chapter 7: The role of quantum error correction and fault-tolerance in the development of large-scale quantum computers.

QUANTUM COMPUTERS HAVE massive potential in the way we process and analyze information. By harnessing the unique properties of quantum mechanics, these devices can solve certain problems much more quickly and efficiently than classical computers. However, the development of large-scale quantum computers has been hindered by a number of challenges, including the fragility of quantum information and the difficulty of controlling quantum systems.

One of the key challenges in building large-scale quantum computers is the inherent fragility of quantum information. Unlike classical bits, which can be either 0 or 1, quantum bits (qubits) can simultaneously exist in a superposition of both states. This allows them to store and process more information than classical bits, but it also makes them extremely sensitive to external perturbations. Any interaction with the environment, such as a stray electromagnetic field or a temperature fluctuation, can cause a quantum system to decohere, losing its quantum properties and becoming classical. This makes it difficult to maintain the fragile quantum states needed for computation.

To overcome this challenge, researchers have developed a number of strategies for protecting and stabilizing quantum information. One of the most important of these is quantum error correction (QEC), a technique that allows quantum systems to maintain their coherence even in the presence of noise and decoherence. By encoding quantum information redundantly across multiple qubits, QEC allows the system to detect and correct errors before they can cause the computation to fail.

However, QEC is not a perfect solution. It requires a large number of qubits to encode each logical qubit, and it is only effective against certain types of errors. To overcome these limitations, researchers have also developed techniques for fault-tolerant quantum computing. Fault-tolerance allows a quantum computer to continue operating even in the presence of faulty components, such as qubits that have decohered or gates that have been improperly applied. Using multiple error correction layers and sophisticated algorithms, fault-tolerant quantum computers can operate with much higher reliability than non-fault-tolerant systems.

The development of quantum error correction and fault-tolerance has been a major milestone in the field of quantum computing. These techniques have allowed researchers to build larger and more complex quantum systems, paving the way for the development of large-scale quantum computers. In recent years, significant progress has been made in the construction of quantum error-correcting codes, with researchers demonstrating the ability to correct errors in small-scale systems using a variety of different approaches.

Despite these advances, there is still much work to be done before large-scale quantum computers become a reality. One of the key challenges is the need for high-quality qubits, which are stable and can be controlled with high precision. Another challenge is the development of scalable quantum algorithms that can take advantage of the unique properties of quantum computers. However, with continued progress in these areas, it is likely that quantum error correction and fault-tolerance will play a crucial role in the eventual realization of large-scale quantum computers.

Chapter 8: The use of quantum simulators to study complex quantum systems and phenomena.

QUANTUM SIMULATORS are a type of computational device that are used to study complex quantum systems and phenomena. Unlike classical computers, which use bits to store and process information, quantum simulators use quantum bits, or qubits, which can exist in multiple states simultaneously. This allows them to simulate the behavior of quantum systems much more accurately than classical computers.

One of the main challenges in the field of quantum mechanics is the study of complex quantum systems. These systems, which can include atoms, molecules, and even larger objects, often have a large number of particles and can be difficult to analyze using classical methods. Quantum simulators, on the other hand, are specifically designed to handle such complex systems and can help researchers better understand the behavior of these systems.

One of the key advantages of quantum simulators is their ability to simulate the behavior of quantum systems under a variety of conditions. This allows researchers to study the effects of different factors, such as temperature, pressure, and external fields, on the behavior of quantum systems. For example, quantum simulators can be used to study the behavior of atoms in a gas, the conductivity of materials, or the behavior of molecules in a chemical reaction.

Another important advantage of quantum simulators is their ability to simulate quantum systems that are difficult or impossible to study using other methods. For example, quantum simulators can be used to study the behavior of particles at very low temperatures, where classical methods are not accurate. They can also be used to study the behavior of particles in extreme environments, such as at very high pressures or in the presence of strong magnetic fields.

In addition to their ability to simulate complex quantum systems, quantum simulators are also valuable for studying the fundamental principles of quantum mechanics. By simulating the behavior of quantum systems, researchers can gain a better understanding of the underlying principles that govern the behavior of these systems. This can help to advance our understanding of the fundamental principles of quantum mechanics and may even lead to the development of new technologies and applications.

One of the key challenges in the use of quantum simulators is the need for highly specialized equipment and expertise. Quantum simulators require a high degree of control over the physical environment in which they operate and sophisticated algorithms and software to perform the simulations. This can make them difficult and expensive to operate, which may limit their use to large research institutions and universities.

Overall, quantum simulators are a valuable tool for studying complex quantum systems and phenomena. They can simulate the behavior of these systems under a wide range of conditions and can provide insight into the fundamental principles of quantum mechanics. While the use of quantum simulators may be limited by the need for specialized equipment and expertise, their potential to advance our understanding of the quantum world makes them an important area of research.

Chapter 9: The applications of quantum computing in cryptography and information security.

QUANTUM COMPUTING IS a rapidly emerging field that can change many areas of science and technology, including cryptography and information security. At its most fundamental level, quantum computing makes use of the principles of quantum mechanics to perform calculations that are beyond the capabilities of classical computers. This has many exciting applications in the field of cryptography, where it can be used to break codes and encrypt information in far more secure ways than what is currently possible with classical computers.

One of the key applications of quantum computing in cryptography is in the area of codebreaking. Classical computers rely on mathematical algorithms to break codes, which can be time-consuming and difficult. Quantum computers, on the other hand, can use the principles of quantum mechanics to quickly and efficiently break codes that are currently considered unbreakable by classical computers. This has the potential to revolutionize the field of codebreaking, making it possible to break codes that were previously thought to be completely secure.

Another important application of quantum computing in cryptography is in the area of encryption. Encryption is the process of encoding information in a way that only authorized individuals can access it. Classical encryption methods are based on mathematical algorithms that can be difficult to break but are not perfect. Quantum computers, on the other hand, can use the principles of quantum mechanics to encrypt information in a virtually impossible way to break. This makes quantum computing a powerful tool for securing sensitive information, such as financial transactions or classified government documents.

In addition to these applications, quantum computing also has the potential to revolutionize other areas of information security. For example, quantum computing can be used to create more secure communication channels, making it difficult for hackers to intercept and decrypt sensitive information. It can also be used to improve the security of computer networks, making it more difficult for attackers to gain access to sensitive information.

The applications of quantum computing in cryptography and information security are vast and exciting. As the field continues to develop and advance, it is likely that we will see even more powerful and sophisticated uses for quantum computers in the realm of information security. This will not only help to protect sensitive information from being accessed by unauthorized individuals, but it will also help to ensure that our digital world remains secure and safe for everyone.

Chapter 10: The potential for quantum computing to revolutionize machine learning and artificial intelligence.

QUANTUM COMPUTING HAS the ability to greatly impact machine learning and artificial intelligence. This is because quantum computers are capable of performing calculations that classical computers would be unable to complete in a reasonable amount of time.

One of the main limitations of classical computers is that they store and process information using bits, which can only be in one of two states: 0 or 1. In contrast, quantum computers use quantum bits, or qubits, which can exist in multiple states simultaneously. This allows quantum computers to perform multiple calculations at the same time, significantly speeding up the process.

Another advantage of quantum computing is that it is based on the principles of quantum mechanics, which govern the behavior of particles on a microscopic level. This means that quantum computers can simulate the behavior of quantum systems, such as molecules, with a high degree of accuracy.

One of the main applications of quantum computing in machine learning is in the field of neural networks. Neural networks are a type of machine learning algorithm that are modeled after the structure of the human brain. They are composed of a large number of interconnected nodes, which are used to process and analyze data.

Quantum computers are well-suited for training and running neural networks, because they can perform the complex calculations required to process large amounts of data. This could enable machine learning algorithms to make more accurate predictions and find patterns in data that would be difficult or impossible to detect using classical computers.

Another potential application of quantum computing in machine learning is in the field of natural language processing (NLP). NLP is a branch of artificial intelligence that focuses on enabling machines to understand and generate human language. Quantum computers could potentially be used to process large amounts of text data and extract meaning from it, which could help improve the accuracy of machine translation and other NLP applications.

The potential of quantum computing to revolutionize machine learning and artificial intelligence is enormous. It has the potential to enable machines to process and analyze vast amounts of data more quickly and accurately than ever before, which could lead to significant advances in fields such as medicine, finance, and logistics. While there are still many challenges to overcome in the development of quantum computers, the potential benefits of this technology are truly exciting.

Chapter 11: The use of quantum computing to study complex systems in fields such as finance and economics.

QUANTUM COMPUTING HAS garnered significant attention in recent years for its ability to handle complex calculations that would be impractical or impossible for classical computers. One area where quantum computing has shown promise is in the study of complex systems in fields such as finance and economics.

Quantum computers use quantum bits, or qubits, to encode and manipulate information. Unlike classical bits, which can only exist in one of two states (0 or 1), qubits can exist simultaneously in a superposition of both states. This allows quantum computers to perform multiple calculations at once, greatly speeding up the computational process.

One application of quantum computing in finance and economics is in the study of optimization problems. These are problems that involve finding the optimal solution among a large number of possible options. For example, a financial company might use an optimization algorithm to determine a portfolio's best investment strategy.

Classical computers can solve optimization problems, but they become inefficient as the number of possible solutions increases. Quantum computers, on the other hand, can solve these problems much more quickly thanks to their ability to perform multiple calculations at once.

Another potential use for quantum computing in finance and economics is in the study of complex networks. These networks, which are common in fields like finance and economics, consist of many interconnected elements, such as stocks in a stock market or individuals in an economy.

Studying these networks can provide valuable insights into the behavior of complex systems. For example, researchers might use network analysis to study the interdependence of different stocks in a stock market, or the interactions between individuals in an economy.

Classical computers can analyze complex networks, but they are limited in their ability to handle the large amounts of data involved. Quantum computers, with their ability to process large amounts of data quickly, could potentially provide a more efficient and effective way of studying these systems.

Quantum computing has the potential to greatly enhance our ability to study complex systems in fields such as finance and economics. By allowing us to quickly solve optimization problems and analyze complex networks, quantum computers could provide valuable insights into the behavior of these systems.

Chapter 12: The potential for quantum computing to revolutionize drug design and the pharmaceutical industry.

QUANTUM COMPUTING HAS the ability to greatly impact the pharmaceutical industry and drug design. This technology has the potential to greatly accelerate the process of discovering and developing new drugs.

One way in which quantum computing can be used in the pharmaceutical industry is by simulating molecular structures. Traditional computers can struggle to simulate complex molecules, but quantum computers can handle these simulations much more efficiently. This could allow researchers to better understand the interactions between different molecules and design more effective drugs.

Another potential use for quantum computing in the pharmaceutical industry is in the area of machine learning. Quantum computers can be used to train machine learning algorithms that can analyze large amounts of data and identify patterns that would be difficult for humans to detect. This could help researchers identify potential new drugs faster and with more accuracy.

Additionally, quantum computing could be used to optimize the production of drugs. By using quantum algorithms, researchers can determine the most efficient methods for synthesizing drugs, reducing the time and cost of production.

The adoption of quantum computing in the pharmaceutical industry has the potential to greatly improve the speed and accuracy of drug discovery and development. This technology could help researchers design new drugs more quickly and efficiently, ultimately benefiting patients and advancing medical science.

Chapter 13: The applications of quantum computing in optimization and combinatorial optimization problems.

QUANTUM COMPUTING IN recent years has received more attention for its ability to tackle complex optimization and combinatorial optimization problems. These problems involve finding the best possible solution from a large number of possibilities, and are commonly found in areas such as scheduling, logistics, and network design.

One of the key advantages of quantum computing in optimization is its ability to process vast amounts of data simultaneously. This is made possible by the unique properties of quantum bits, or qubits, which can exist in multiple states simultaneously. This allows a quantum computer to explore multiple solutions to a problem at the same time, making it possible to find the optimal solution much faster than with classical computing methods.

Another advantage of quantum computing in optimization is its ability to quickly search large spaces of possibilities. This is known as "quantum speedup," and it is made possible by a quantum algorithm called Grover's algorithm. This algorithm allows a quantum computer to search an unsorted database quadratically faster than classical methods, making it a powerful tool for solving optimization problems.

In addition to optimization, quantum computing also has the ability to solve combinatorial optimization problems. Combinatorial optimization involves finding the optimal arrangement of items from a given set of possibilities. Examples of combinatorial optimization problems include the traveling salesman person problem, in which the goal is to find the shortest possible route that visits a set of cities, and the knapsack problem, in which the goal is to maximize the value of items that can be placed in a knapsack of a given size.

Quantum computers can solve these problems using a technique called quantum annealing. This involves encoding the problem into the energy levels of a quantum system, and then allowing the system to evolve naturally towards the lowest energy state. This approach has been shown to be very effective for solving combinatorial optimization problems, and could have many applications in fields such as logistics and supply chain management.

Quantum computing has the potential to significantly improve our ability to solve complex optimization and combinatorial optimization problems. It has the potential to enable faster, more accurate solutions to these problems, which could have a major impact on many areas of science and industry.

Chapter 14: The role of quantum computing in the study of quantum field theories and high-energy physics.

QUANTUM COMPUTING HAS the ability to understand the universe and the fundamental laws that govern it. One area where it is particularly promising is in the study of quantum field theories and high-energy physics.

A quantum field theory is a theoretical framework that describes the behavior of subatomic particles, such as electrons and quarks, in terms of fields that are defined over all of space-time. These theories are essential to our understanding of the fundamental forces of nature, such as electromagnetism and the strong and weak nuclear forces.

One of the key challenges in studying quantum field theories is the required immense computational power. This is because these theories involve an enormous number of particles and interactions, and traditional computers are not well-suited to simulating these complex systems.

This is where quantum computing comes in. Quantum computers are able to perform calculations that are exponentially faster than those that are possible on classical computers. This means that they are ideally suited to studying the complex systems that arise in quantum field theories and high-energy physics.

One example of how quantum computing is being used in this context is in the study of lattice gauge theories. These are theories that describe the behavior of particles on a lattice, which is a regular array of points in space-time. Simulating these theories on a classical computer is extremely difficult, but quantum computers are able to do it much more efficiently.

Another example is in the study of quantum chromodynamics, which is the theory of the strong nuclear force. This force is responsible for binding quarks together to form protons and neutrons, which is one of nature's fundamental forces. Simulating quantum chromodynamics on a classical computer is extremely challenging, but quantum computers are able to do it much more efficiently.

The role of quantum computing in the study of quantum field theories and high-energy physics is incredibly important. It is providing us with new tools and insights that are helping us to understand the fundamental laws of the universe better. As quantum computing continues to advance, it is likely that it will continue to play a vital role in this field for many years to come.

Chapter 15: The potential for quantum computing to revolutionize the study of quantum gravity and cosmology.

QUANTUM COMPUTING CAN be utilized in the study of quantum gravity and cosmology. This is because quantum computers operate on the principles of quantum mechanics, which govern the behavior of subatomic particles and are essential to understanding the fundamental nature of the universe.

One of the biggest challenges in studying quantum gravity and cosmology is the computational complexity of the calculations involved. These calculations often require massive amounts of computing power, making them difficult or even impossible to carry out on classical computers. Quantum computers, on the other hand, are capable of performing these calculations much more efficiently due to their unique ability to process and manipulate large amounts of data simultaneously.

For example, simulating the behavior of black holes and other extreme gravitational phenomena is a key goal of quantum gravity research. Black holes are incredibly complex objects, with their behavior governed by general relativity and quantum mechanics principles. Simulating the behavior of a black hole on a classical computer would require an enormous amount of computational power, but a quantum computer could potentially do it much more efficiently.

Similarly, cosmology studies the origins and evolution of the universe, a field heavily reliant on mathematical modeling and simulations. Quantum computers could be used to carry out these simulations much more quickly and accurately than classical computers, allowing scientists to make more precise predictions and better understand the fundamental nature of the universe.

In addition to their potential use in simulating complex phenomena, quantum computers could also be used to study the fundamental nature of space and time. This is because quantum mechanics allows for the manipulation of individual particles at the subatomic level, which could potentially be used to probe the structure of space and time at the most fundamental level. This could provide new insights into the fundamental nature of the universe, and could even lead to the development of new theories of quantum gravity.

The potential for quantum computing to revolutionize the study of quantum gravity and cosmology is vast. By enabling more efficient and accurate simulations of complex phenomena, and by providing new tools for studying the fundamental nature of the universe, quantum computers could help unlock many of the mysteries of the cosmos.

Chapter 16: The role of quantum computing in the study of complex systems and emergent phenomena.

QUANTUM COMPUTING CAN be integrated with the study of complex systems and emergent phenomena. This is because quantum computers are able to perform certain types of calculations much faster than classical computers, which makes them well-suited for analyzing complex systems.

One of the key challenges in the study of complex systems is the so-called "combinatorial explosion" - the rapid increase in the number of possible states or configurations that a system can have as its size increases. This makes it difficult to analyze such systems using classical computers, which can only process a limited amount of information at a time.

Quantum computers, on the other hand, are able to process vast amounts of information simultaneously, using a phenomenon known as quantum parallelism. This allows them to efficiently explore the vast space of possible configurations of a complex system, and to identify patterns and correlations that would be impossible to detect using classical methods.

For example, quantum computers could be used to study the behavior of large networks of interacting particles, such as those found in materials science or biology. By simulating the quantum interactions between these particles, researchers could gain insights into the system's emergent properties as a whole, such as its conductivity or ability to transport energy.

Similarly, quantum computers could be used to study the behavior of large social networks, such as those found on social media platforms. By analyzing the patterns of interactions between individuals in these networks, researchers could gain insights into the emergence of collective behaviors, such as the spread of information or the formation of opinion clusters.

With their ability to analyze complex systems, quantum computers also have the potential to enhance our understanding of emergent phenomena. This is because many emergent phenomena arise from the collective behavior of large numbers of individual components, which can only be fully understood by examining the interactions between these components at a very fine level of detail.

For example, quantum computers could be used to study the emergence of collective intelligence in social networks, by simulating the interactions between individual agents and examining how these interactions give rise to collective decision-making and problem-solving abilities.

The overall ability of quantum computers to efficiently explore large spaces of possible configurations, and to analyze the interactions between individual components at a very fine level of detail, makes them uniquely suited for studying complex systems and emergent phenomena. As quantum computing technology continues to advance, it is likely that it will play an increasingly important role in advancing our understanding of these fascinating phenomena.

Chapter 17: The potential for quantum computing to advance our understanding of the fundamental nature of reality.

QUANTUM COMPUTING CAN better our understanding of the fundamental nature of reality. At its most basic level, quantum computing relies on the principles of quantum mechanics, which describe the behavior of particles on a tiny, atomic scale. This allows quantum computers to operate in ways that are fundamentally different from classical computers, which are based on the laws of classical mechanics.

One of the key differences between classical and quantum computing is the way in which they process information. Classical computers use bits, which can only represent information as either a 0 or a 1. Quantum computers, on the other hand, use qubits, which can represent information as both a 0 and a 1 simultaneously. This ability to be in multiple states at once is known as superposition, and it allows quantum computers to perform certain calculations much more quickly than classical computers.

One potential application of quantum computing is in the field of particle physics. By simulating the behavior of particles at the quantum level, quantum computers could help scientists understand the fundamental forces that govern the universe. This could lead to a deeper understanding of reality's nature and help researchers develop new technologies and materials.

Another potential use for quantum computing is in the field of chemistry. By simulating the behavior of molecules at the quantum level, quantum computers could help researchers understand the complex reactions that take place in chemical reactions. This could lead to developing new drugs and other chemical compounds, which could significantly benefit society.

Additionally, quantum computing could have applications in the field of cryptography. Because of the inherent security of quantum mechanics, quantum computers could be used to encrypt information in ways that are impossible to crack using classical computers. This could have major implications for online security and could help protect sensitive information from being accessed by unauthorized individuals.

The potential for quantum computing to advance our understanding of the fundamental nature of reality is vast. By utilizing the principles of quantum mechanics, quantum computers could help researchers simulate and understand the behavior of particles and molecules at the quantum level. This could lead to new discoveries and technological developments in fields ranging from particle physics to chemistry and cryptography. While there are still many challenges to overcome before quantum computing becomes a reality, the potential benefits are enormous and could have a profound impact on our understanding of the universe and our place in it.

Chapter 18: The applications of quantum computing in finance include using quantum algorithms for portfolio optimization and risk management.

ONE OF THE PRIMARY ways that quantum computing could be applied in finance is through the use of quantum algorithms for portfolio optimization and risk management.

Portfolio optimization is the process of selecting a mix of investments that will maximize returns while minimizing risk. This is a complex task, especially for large portfolios with many different assets. Traditional computers can handle this task, but they often struggle to find the optimal solution in a reasonable amount of time.

Quantum computers, on the other hand, have the potential to solve these types of problems much more quickly. This is because quantum computers use quantum bits, or qubits, which can exist in multiple states simultaneously. This allows them to process a huge number of possibilities at once, which makes them well-suited to optimization problems.

In the context of portfolio optimization, a quantum computer could be used to quickly evaluate a large number of potential investment portfolios and determine the best one based on a set of predefined criteria. This could be done in a fraction of the time it would take a traditional computer, allowing investment managers to make more informed decisions and potentially improve the performance of their portfolios.

Risk management is another important area where quantum computing could have a significant impact. Risk management is the process of identifying, analyzing, and mitigating potential risks that could affect a financial institution or investment portfolio. This is a critical task in finance, as it helps to ensure that an institution is prepared for unexpected events and can minimize potential losses.

Traditionally, risk management has relied on statistical analysis and mathematical modeling to identify and evaluate risks. However, these methods can be limited by the amount of data that is available and the complexity of the models. Quantum computers, with their ability to process vast amounts of data quickly, could be used to develop more sophisticated models and perform more accurate risk assessments.

In summary, the applications of quantum computing in finance include using quantum algorithms for portfolio optimization and risk management. These algorithms could help investment managers make more informed decisions and improve the performance of their portfolios, and they could also be used to develop more sophisticated models for risk assessment. While the use of quantum computing in finance is still in its early stages, it can revolutionize the industry and provide significant benefits for investors and financial institutions.

Chapter 19: The potential for quantum computing to revolutionize the field of astronomy and the search for extraterrestrial life.

QUANTUM COMPUTERS ARE capable of processing vast amounts of data at incredibly high speeds, making them ideal for analyzing the massive amounts of data generated by telescopes and other astronomical instruments.

One way that quantum computing could be used in astronomy is to help search for signs of extraterrestrial life. Currently, the search for life beyond Earth is limited by the amount of data that existing computers can process. However, with quantum computers, it would be possible to analyze much larger datasets in a shorter amount of time, increasing the chances of finding evidence of extraterrestrial life.

Another potential application of quantum computing in astronomy is in the study of exoplanets. Exoplanets are planets that orbit stars other than our Sun, and the number of known exoplanets has been increasing rapidly in recent years. However, many of these exoplanets are difficult to study because they are too far away or too small to be observed directly. Quantum computers could be used to analyze the data collected by telescopes and other instruments to learn more about these distant worlds.

Quantum computing could also be used to improve our understanding of the universe as a whole. For example, it could be used to simulate the behavior of complex astrophysical phenomena, such as the formation of galaxies or the evolution of stars. This could help astronomers to better understand the processes that shape the universe and the conditions under which life might arise.

The potential for quantum computing to revolutionize the field of astronomy and the search for extraterrestrial life is immense. By allowing us to process vast amounts of data quickly and accurately, quantum computers could help us to unlock many of the mysteries of the universe and to search for signs of life beyond our own planet.

Chapter 20: The future of quantum computing, including potential breakthroughs and developments in the field.

QUANTUM COMPUTING IS a rapidly developing field that holds great potential for many applications, including improving artificial intelligence, speeding up drug discovery, and enabling secure communication. In the future, we can expect to see many exciting breakthroughs and developments in this field as researchers continue to push the boundaries of what is possible.

One of the most significant potential breakthroughs in quantum computing is the development of a large-scale quantum computer. While quantum computers currently exist, they are limited in their capabilities and only able to solve certain problems. A large-scale quantum computer, on the other hand, would be able to tackle a much wider range of problems, making it a highly valuable tool for many different industries.

Another potential development in the field of quantum computing is the creation of robust quantum algorithms. Quantum algorithms are used to solve specific problems using quantum computers, and their development is essential for the practical application of quantum computers. In the future, we can expect to see many new and improved quantum algorithms that will enable quantum computers to solve even more complex problems.

Another exciting development in the field of quantum computing is the integration of quantum technology with classical computing. This would involve using quantum computers to perform specific tasks that are currently too difficult or time-consuming for classical computers, while still being able to use classical computers for other tasks. This could lead to a significant increase in the overall efficiency and speed of computation.

Additionally, we can expect to see further advances in the field of quantum communication, which is essential for securing information transmission. Quantum communication uses the principles of quantum mechanics to transmit information in a way that is secure from tampering or interception. In the future, we can expect to see the development of more advanced quantum communication systems that are faster, more secure, and more efficient.

Overall, the future of quantum computing is incredibly promising, and we can expect to see many exciting breakthroughs and developments in the field. As quantum computers continue to evolve, they will become an increasingly valuable tool for solving complex problems and improving many different industries.

Chapter 21: The future of quantum mechanics, including potential breakthroughs and developments in the field.

QUANTUM MECHANICS IS a fundamental theory of physics that describes the behavior of matter and energy at the atomic and subatomic level. It is a powerful and successful theory that experiments have widely tested and confirmed. In the future, quantum mechanics will likely continue to be a major focus of research in physics and other fields, and several potential breakthroughs and developments could arise in the field.

One potential development in the field of quantum mechanics is the creation of large-scale quantum computers. These computers would use the principles of quantum mechanics to perform calculations that are currently beyond the capabilities of classical computers. This could lead to major advances in fields such as cryptography, materials science, and drug design.

Another potential breakthrough in quantum mechanics could be the discovery of new phenomena or particles. For example, physicists have recently observed hints of a new type of subatomic particle called the axion, which could help explain some of the mysteries of dark matter. The discovery of new particles or phenomena would deepen our understanding of the fundamental nature of the universe.

In addition to these potential developments, several ongoing research efforts in quantum mechanics could lead to significant advances in the field. For example, physicists are working on developing new technologies for measuring and controlling quantum systems, such as ion traps and superconducting circuits. These technologies could enable more precise experiments and more complex quantum systems.

Overall, the future of quantum mechanics is likely to be marked by continued research and experimentation, with the potential for major breakthroughs and developments. As our understanding of quantum mechanics deepens, it will likely have important implications for a wide range of fields, from computing and communication to materials science and medicine.

Chapter 22: Continued development and improvement of quantum computers, leading to more advanced and powerful systems.

QUANTUM COMPUTERS ARE a type of computer that uses the principles of quantum mechanics to perform calculations and process information. Unlike traditional computers, which use bits that are either 0 or 1, quantum computers use quantum bits, or qubits, which can be 0, 1, or a superposition of both 0 and 1. This allows quantum computers to perform certain types of calculations much faster than classical computers.

Over the past few years, there has been a significant amount of progress in the field of quantum computing. Scientists and engineers have developed more advanced and powerful quantum computer systems, and they are continuing to work on improving these systems. Some of the key areas of focus in this ongoing development include increasing the number of qubits, improving qubit coherence and control, and developing new algorithms and software that can take advantage of the unique capabilities of quantum computers.

One of the main challenges in building quantum computers is the difficulty of maintaining coherence in the qubits. Coherence refers to the ability of the qubits to maintain their quantum state, which is essential for performing calculations. To overcome this challenge, researchers are working on developing new techniques for controlling and manipulating qubits, such as using superconducting materials and advanced cooling methods.

Another key area of focus is the development of new algorithms and software that can exploit the unique capabilities of quantum computers. Classical computers can only process one piece of data at a time, whereas quantum computers can process multiple pieces of data simultaneously. This allows quantum computers to perform certain types of calculations much faster than classical computers, which has the potential to revolutionize fields such as cryptography, drug discovery, and optimization.

In conclusion, the continued development and improvement of quantum computers is leading to more advanced and powerful systems. By overcoming challenges such as qubit coherence and control, and developing new algorithms and software, researchers are paving the way for the widespread adoption of quantum computers in a variety of fields.

Chapter 23: The integration of quantum computing into a wider range of fields and industries, such as finance, medicine, and materials science.

QUANTUM COMPUTING IS a cutting-edge technology that uses the principles of quantum mechanics to perform calculations that are significantly faster and more powerful than those performed by traditional computers. Over the past few years, researchers have been working to develop and improve quantum computing technology, and it is now starting to be integrated into a wider range of fields and industries.

One of the areas where quantum computing is starting to have a major impact is finance. In the world of finance, complex calculations and simulations are performed regularly, and the ability to do these calculations quickly and accurately is essential. Quantum computers can perform these calculations much faster than traditional computers, which allows financial institutions to make more informed and accurate predictions about market trends and investment opportunities.

Another area where quantum computing is starting to be used is medicine. Quantum computers can be used to help researchers analyze vast amounts of data, such as genetic data from large populations, to identify patterns and correlations that would be difficult or impossible to detect using traditional computers. This can help doctors and researchers develop more effective treatments for a wide range of medical conditions.

Additionally, quantum computing is starting to be used in materials science. The ability of quantum computers to perform complex calculations quickly and accurately is proving to be extremely useful in the development of new materials and chemicals. For example, quantum computers can be used to simulate the behavior of atoms and molecules, which can help researchers design new materials with specific properties, such as increased strength or improved conductivity.

Overall, the integration of quantum computing into a wider range of fields and industries is a promising development that has the potential to revolutionize many different fields. As quantum computing technology continues to advance, we can expect to see even more exciting applications of this technology in the future.

Chapter 24: The emergence of new quantum algorithms and applications, which could unlock previously inaccessible computational tasks.

IN RECENT YEARS, THERE has been a surge of interest in the development of new quantum algorithms and applications. These algorithms make use of the unique properties of quantum mechanics to solve computational problems that are difficult or impossible for classical computers to handle.

One of the most well-known examples of a quantum algorithm is Shor's algorithm, which can be used to factorize large numbers efficiently. This has important implications for cryptography, as many commonly used encryption methods rely on the difficulty of factoring large numbers. By using Shor's algorithm, a quantum computer could potentially break these encryption schemes, making them vulnerable to attack.

Another area where quantum algorithms have shown promise is in machine learning. Quantum machine learning algorithms have the potential to process vast amounts of data much more quickly than classical algorithms, making them well-suited for tasks such as pattern recognition and data classification.

In addition to these applications, researchers are also exploring the use of quantum algorithms for a wide range of other tasks, including optimization, simulation, and even the design of new materials.

While the development of quantum algorithms is still in its early stages, it is clear that they have the potential to unlock previously inaccessible computational tasks. This could have far-reaching consequences for many fields, from cryptography and finance, to medicine and science. As research in this area continues to advance, we can expect to see even more exciting developments in the future.

Chapter 25: The exploration of new forms of quantum mechanics, such as topological quantum computing and quantum annealing, which could provide even more powerful computing capabilities.

QUANTUM MECHANICS IS a branch of physics that deals with the behavior of particles on a microscopic scale, such as atoms and subatomic particles. In recent years, there has been a lot of interest in exploring new forms of quantum mechanics that could potentially lead to even more powerful computing capabilities.

One area of research that has gained a lot of attention is the field of topological quantum computing. In traditional computing, information is stored in bits, which can be either a 0 or a 1. In contrast, topological quantum computing uses qubits, which can be in a superposition of both 0 and 1 at the same time. This allows for much more complex calculations to be performed in parallel, potentially leading to a much faster and more efficient form of computing.

Another area of research is quantum annealing, which uses quantum mechanics to find the optimal solution to a problem. This is done by encoding the problem into the energy levels of a system, and then using quantum effects to find the lowest energy state, which corresponds to the optimal solution. Quantum annealing has the potential to be used in a wide range of applications, including machine learning and optimization problems.

While these new forms of quantum mechanics are still in the early stages of development, they hold great promise for the future of computing. By harnessing the power of quantum mechanics, we may be able to build computers that can solve problems that are currently impossible for classical computers to handle.

Chapter 26: Collaboration and partnerships between academia, industry, and government to advance the field of quantum computing and mechanics.

COLLABORATION AND PARTNERSHIPS between academia, industry, and government have played a crucial role in advancing the field of quantum computing and mechanics. Academia has provided the foundational knowledge and expertise in quantum mechanics and computing, while industry has helped to develop and commercialize this technology. Government agencies, on the other hand, have provided funding and support for research and development in this area.

One example of a successful collaboration between academia, industry, and government is the National Institute of Standards and Technology (NIST) and the University of Colorado Boulder's Joint Quantum Institute (JQI). The JQI is a partnership between NIST and the university that aims to advance the field of quantum information science. The institute brings together researchers from both academia and industry to conduct research and development in areas such as quantum computing, quantum communication, and quantum metrology.

Another example is the partnership between the University of Maryland and Google, which has resulted in the development of a quantum computer called Bristlecone. This collaboration has brought together the expertise of academia in the field of quantum mechanics and the resources of a major technology company to advance the development of quantum computing.

Overall, collaboration and partnerships between academia, industry, and government have been instrumental in driving the growth and development of the field of quantum computing and mechanics. By combining their expertise and resources, these groups have been able to make significant advances in this field and lay the groundwork for future breakthroughs.

Chapter 27: The Future Is Bright for Quantum Computing & Mechanics.

THE FUTURE IS BRIGHT for quantum computing and mechanics because these fields have the potential to revolutionize the way we process information and understand the fundamental nature of the universe. In recent years, there have been significant advances in the development of quantum computers, which are capable of performing calculations that are exponentially faster than those of classical computers. This could potentially lead to breakthroughs in a wide range of fields, including cryptography, pharmaceuticals, and artificial intelligence.

Similarly, the study of quantum mechanics, which is the theory that describes the behavior of particles on a quantum level, has also made significant progress in recent years. This has led to a better understanding of the fundamental nature of the universe, and has given rise to new technologies such as quantum sensors and quantum teleportation.

There are still many challenges and obstacles to overcome in both of these fields, but the potential rewards are so great that many scientists and researchers are committed to pushing the boundaries of what is possible. As a result, it is likely that we will continue to see rapid progress and exciting developments in the years to come.

Conclusion

IN CONCLUSION, QUANTUM computing and quantum mechanics are fascinating fields of study that hold immense potential for revolutionizing various industries and advancing our understanding of the universe. Quantum computers are based on the principles of quantum mechanics, which is a branch of physics that deals with the behavior of subatomic particles. Unlike classical computers, which are based on binary bits that can only represent 0s or 1s, quantum computers use quantum bits, or qubits, which can represent 0s, 1s, or both at the same time. This allows quantum computers to perform calculations much faster than classical computers.

One of the most promising applications of quantum computing is in the field of cryptography. Since quantum computers can solve certain mathematical problems much faster than classical computers, they can be used to break many of the encryption algorithms that are currently used to secure sensitive information. This has led to the development of quantum-resistant cryptography, which uses mathematical problems that are believed to be beyond the capabilities of even the most powerful quantum computers.

Another important application of quantum computing is in the field of chemistry. By simulating the behavior of atoms and molecules, quantum computers can help chemists design new drugs and materials, and understand the mechanisms of chemical reactions. This could lead to the development of more efficient and environmentally friendly chemicals, as well as a better understanding of complex biological systems.

Regarding their practical applications, quantum computers and quantum mechanics also have important implications for our understanding of the fundamental nature of the universe. Quantum mechanics has challenged our classical concepts of space and time, and has led to the development of theories such as quantum entanglement and the uncertainty principle. These theories suggest that the universe is fundamentally different from what we experience in our everyday lives, and that our classical ideas of cause and effect may not always hold true at the quantum level.

While there is still much to learn about quantum computing and quantum mechanics, the potential benefits and insights that they offer are truly exciting.

As our understanding of these fields continues to grow, we can expect to see more and more practical applications, as well as new breakthroughs in our understanding of the fundamental nature of the universe.

Don't miss out!

Visit the website below and you can sign up to receive emails whenever Adam Smith publishes a new book. There's no charge and no obligation.

https://books2read.com/r/B-A-OSAW-HYNDC

BOOKS 2 READ

Connecting independent readers to independent writers.

Did you love *The Beginner's Guide to Quantum Computing & Mechanics*? Then you should read *40 Things Intelligent People Do Daily* by Adam Smith!

"40 Things Intelligent People Do Daily" is a hypothetical book that explores the daily habits and routines of intelligent people. The book examines how these individuals approach tasks and challenges, and how they use their intelligence to improve their lives and achieve their goals. The book provides practical tips and advice for readers looking to develop their own intelligent habits and become more successful in their personal and professional lives.

Read more at https://books2read.com/adamsmith.

About the Author

Adam Smith is a native North Texas Entrepreneur, social media expert, and educator who graduated from Embry Riddle University with a Bachelor of Science in Aerospace Studies. In 2007 he co-founded TeacherTube.com, an online education platform community that provides an educationally focused, safe venue for teachers, schools, and home learners. Under his leadership, in September of 2017, a publicly traded company, Salem Media Group, acquired TeacherTube. At the time of acquisition, TeacherTube had over 2 million registered educators with over 12 million monthly unique views of the site.

After the acquisition, Adam continued his passion for technology and education by helping schools and start-ups to large corporations with online training, branding, content marketing, and social media management.

To find out more about Adam, go to: https://books2read.com/adamsmith
Read more at https://books2read.com/adamsmith.

About the Publisher

A. Smith Media is a fully integrated digital media and distribution company specializing in producing and marketing television, film, home entertainment, music, and interactive games. A. Smith Media has been at the forefront of the commercial growth of Digital Distribution and Internet Video on Demand services through its pioneering efforts in web TV.